# M

MW01206189

| MAKE: | | M | | |
|-------|---|---|---|---|
| DATE: | ODOMETER. START \| END | | TOTAL. | DESTINATION / PURPOSE: |
| | | | | |
| | | | | |
| | | | | |
| | | | | |
| | | | | |
| | | | | |
| | | | | |
| | | | | |
| | | | | |
| | | | | |
| | | | | |
| | | | | |
| | | | | |

# MILEAGE LOG

| MAKE: | | MODEL: | | YEAR: |
|---|---|---|---|---|
| **DATE:** | **ODOMETER: START \| END** | | **TOTAL:** | **DESTINATION / PURPOSE:** |
| | | | | |
| | | | | |
| | | | | |
| | | | | |
| | | | | |
| | | | | |
| | | | | |
| | | | | |
| | | | | |
| | | | | |
| | | | | |
| | | | | |
| | | | | |
| | | | | |
| | | | | |
| | | | | |

# MILEAGE LOG

| MAKE: | MODEL: | | YEAR: | |
|---|---|---|---|---|
| DATE: | ODOMETER:<br>START \| END | | TOTAL: | DESTINATION /<br>PURPOSE: |
| | | | | |
| | | | | |
| | | | | |
| | | | | |
| | | | | |
| | | | | |
| | | | | |
| | | | | |
| | | | | |
| | | | | |
| | | | | |
| | | | | |
| | | | | |

# MILEAGE LOG

| MAKE: | | MODEL: | | YEAR: |
|---|---|---|---|---|
| DATE: | ODOMETER: START \| END | | TOTAL: | DESTINATION / PURPOSE: |
| | | | | |
| | | | | |
| | | | | |
| | | | | |
| | | | | |
| | | | | |
| | | | | |
| | | | | |
| | | | | |
| | | | | |
| | | | | |
| | | | | |
| | | | | |

# MILEAGE LOG

| MAKE: | | MODEL: | | YEAR: |
|---|---|---|---|---|
| **DATE:** | **ODOMETER:**<br>**START \| END** | | **TOTAL:** | **DESTINATION /**<br>**PURPOSE:** |
| | | | | |
| | | | | |
| | | | | |
| | | | | |
| | | | | |
| | | | | |
| | | | | |
| | | | | |
| | | | | |
| | | | | |
| | | | | |
| | | | | |
| | | | | |

# MILEAGE LOG

| MAKE: | MODEL: | | YEAR: | |
|-------|--------|---|-------|---|
| DATE: | ODOMETER: START \| END | | TOTAL: | DESTINATION / PURPOSE: |
| | | | | |
| | | | | |
| | | | | |
| | | | | |
| | | | | |
| | | | | |
| | | | | |
| | | | | |
| | | | | |
| | | | | |
| | | | | |
| | | | | |
| | | | | |
| | | | | |

# MILEAGE LOG

| MAKE: | | MODEL: | | YEAR: |
|---|---|---|---|---|
| **DATE:** | **ODOMETER:**<br>**START \| END** | | **TOTAL:** | **DESTINATION /**<br>**PURPOSE:** |
| | | | | |
| | | | | |
| | | | | |
| | | | | |
| | | | | |
| | | | | |
| | | | | |
| | | | | |
| | | | | |
| | | | | |
| | | | | |
| | | | | |

# MILEAGE LOG

| MAKE: | MODEL: | | YEAR: |
|---|---|---|---|
| DATE: | ODOMETER:<br>START \| END | TOTAL: | DESTINATION /<br>PURPOSE: |
| | | | |
| | | | |
| | | | |
| | | | |
| | | | |
| | | | |
| | | | |
| | | | |
| | | | |
| | | | |
| | | | |
| | | | |
| | | | |
| | | | |

# MILEAGE LOG

| MAKE: | | MODEL: | | YEAR: |
|-------|--|--------|--|-------|
| **DATE:** | **ODOMETER:** **START \| END** | | **TOTAL:** | **DESTINATION / PURPOSE:** |
| | | | | |
| | | | | |
| | | | | |
| | | | | |
| | | | | |
| | | | | |
| | | | | |
| | | | | |
| | | | | |
| | | | | |
| | | | | |
| | | | | |
| | | | | |
| | | | | |
| | | | | |
| | | | | |

# MILEAGE LOG

| MAKE: | MODEL: | | YEAR: | |
|-------|--------|--|-------|--|
| DATE: | ODOMETER:<br>START \| END | | TOTAL: | DESTINATION /<br>PURPOSE: |
| | | | | |
| | | | | |
| | | | | |
| | | | | |
| | | | | |
| | | | | |
| | | | | |
| | | | | |
| | | | | |
| | | | | |
| | | | | |
| | | | | |
| | | | | |
| | | | | |

# MILEAGE LOG

| MAKE: | | MODEL: | | YEAR: | |
|---|---|---|---|---|---|
| **DATE:** | **ODOMETER:** **START \| END** | | **TOTAL:** | **DESTINATION /** **PURPOSE:** | |
| | | | | | |
| | | | | | |
| | | | | | |
| | | | | | |
| | | | | | |
| | | | | | |
| | | | | | |
| | | | | | |
| | | | | | |
| | | | | | |
| | | | | | |
| | | | | | |
| | | | | | |
| | | | | | |

# MILEAGE LOG

| MAKE: | | MODEL: | | YEAR: |
|---|---|---|---|---|
| DATE: | ODOMETER: START \| END | | TOTAL: | DESTINATION / PURPOSE: |
| | | | | |
| | | | | |
| | | | | |
| | | | | |
| | | | | |
| | | | | |
| | | | | |
| | | | | |
| | | | | |
| | | | | |
| | | | | |
| | | | | |
| | | | | |
| | | | | |

# MILEAGE LOG

| MAKE: | | MODEL: | | YEAR: |
|---|---|---|---|---|
| DATE: | ODOMETER: START \| END | | TOTAL: | DESTINATION / PURPOSE: |
| | | | | |
| | | | | |
| | | | | |
| | | | | |
| | | | | |
| | | | | |
| | | | | |
| | | | | |
| | | | | |
| | | | | |
| | | | | |
| | | | | |
| | | | | |
| | | | | |

# MILEAGE LOG

| MAKE: | | MODEL: | | YEAR: |
|-------|--|--------|--|-------|
| DATE: | ODOMETER:<br>START \| END | | TOTAL: | DESTINATION /<br>PURPOSE: |
| | | | | |
| | | | | |
| | | | | |
| | | | | |
| | | | | |
| | | | | |
| | | | | |
| | | | | |
| | | | | |
| | | | | |
| | | | | |
| | | | | |
| | | | | |
| | | | | |

# MILEAGE LOG

| MAKE: | | MODEL: | | YEAR: |
|---|---|---|---|---|
| **DATE:** | **ODOMETER:** **START \| END** | | **TOTAL:** | **DESTINATION / PURPOSE:** |
| | | | | |
| | | | | |
| | | | | |
| | | | | |
| | | | | |
| | | | | |
| | | | | |
| | | | | |
| | | | | |
| | | | | |
| | | | | |
| | | | | |
| | | | | |
| | | | | |

# MILEAGE LOG

| MAKE: | MODEL: | | YEAR: | |
|---|---|---|---|---|
| DATE: | ODOMETER:<br>START \| END | | TOTAL: | DESTINATION /<br>PURPOSE: |
| | | | | |
| | | | | |
| | | | | |
| | | | | |
| | | | | |
| | | | | |
| | | | | |
| | | | | |
| | | | | |
| | | | | |
| | | | | |
| | | | | |
| | | | | |

# MILEAGE LOG

| MAKE: | | MODEL: | | YEAR: |
|---|---|---|---|---|
| **DATE:** | **ODOMETER:**<br>**START \| END** | | **TOTAL:** | **DESTINATION /**<br>**PURPOSE:** |
| | | | | |
| | | | | |
| | | | | |
| | | | | |
| | | | | |
| | | | | |
| | | | | |
| | | | | |
| | | | | |
| | | | | |
| | | | | |
| | | | | |
| | | | | |
| | | | | |

# MILEAGE LOG

| DATE: | ODOMETER: START \| END | | TOTAL: | DESTINATION / PURPOSE: |
|-------|---------|---------|--------|------------------------|
| MAKE: | | MODEL: | | YEAR: |
| | | | | |
| | | | | |
| | | | | |
| | | | | |
| | | | | |
| | | | | |
| | | | | |
| | | | | |
| | | | | |
| | | | | |
| | | | | |
| | | | | |
| | | | | |

# MILEAGE LOG

| MAKE: | | MODEL: | | YEAR: |
|---|---|---|---|---|
| DATE: | ODOMETER: START \| END | | TOTAL: | DESTINATION / PURPOSE: |
| | | | | |
| | | | | |
| | | | | |
| | | | | |
| | | | | |
| | | | | |
| | | | | |
| | | | | |
| | | | | |
| | | | | |
| | | | | |
| | | | | |
| | | | | |
| | | | | |

# MILEAGE LOG

| MAKE: | | MODEL: | | YEAR: |
|---|---|---|---|---|
| DATE: | ODOMETER:<br>START \| END | | TOTAL: | DESTINATION /<br>PURPOSE: |
| | | | | |
| | | | | |
| | | | | |
| | | | | |
| | | | | |
| | | | | |
| | | | | |
| | | | | |
| | | | | |
| | | | | |
| | | | | |
| | | | | |
| | | | | |
| | | | | |
| | | | | |

# MILEAGE LOG

| MAKE: | | MODEL: | | YEAR: |
|---|---|---|---|---|
| **DATE:** | **ODOMETER: START \| END** | | **TOTAL:** | **DESTINATION / PURPOSE:** |
| | | | | |
| | | | | |
| | | | | |
| | | | | |
| | | | | |
| | | | | |
| | | | | |
| | | | | |
| | | | | |
| | | | | |
| | | | | |
| | | | | |
| | | | | |
| | | | | |

# MILEAGE LOG

| MAKE: | | MODEL: | | YEAR: |
|---|---|---|---|---|
| **DATE:** | **ODOMETER:**<br>**START \| END** | | **TOTAL:** | **DESTINATION /**<br>**PURPOSE:** |
| | | | | |
| | | | | |
| | | | | |
| | | | | |
| | | | | |
| | | | | |
| | | | | |
| | | | | |
| | | | | |
| | | | | |
| | | | | |
| | | | | |
| | | | | |
| | | | | |

# MILEAGE LOG

| MAKE: | | MODEL: | | YEAR: |
|---|---|---|---|---|
| **DATE:** | **ODOMETER:**<br>**START \| END** | | **TOTAL:** | **DESTINATION /**<br>**PURPOSE:** |
| | | | | |
| | | | | |
| | | | | |
| | | | | |
| | | | | |
| | | | | |
| | | | | |
| | | | | |
| | | | | |
| | | | | |
| | | | | |
| | | | | |
| | | | | |
| | | | | |

# MILEAGE LOG

| MAKE: | MODEL: | | YEAR: |
|---|---|---|---|
| DATE: | ODOMETER:<br>START \| END | TOTAL: | DESTINATION /<br>PURPOSE: |
| | | | |
| | | | |
| | | | |
| | | | |
| | | | |
| | | | |
| | | | |
| | | | |
| | | | |
| | | | |
| | | | |
| | | | |
| | | | |
| | | | |

# MILEAGE LOG

| MAKE: | | MODEL: | | YEAR: |
|---|---|---|---|---|

| DATE: | ODOMETER:<br>START \| END | | TOTAL: | DESTINATION /<br>PURPOSE: |
|---|---|---|---|---|
| | | | | |
| | | | | |
| | | | | |
| | | | | |
| | | | | |
| | | | | |
| | | | | |
| | | | | |
| | | | | |
| | | | | |
| | | | | |
| | | | | |
| | | | | |
| | | | | |

# MILEAGE LOG

| MAKE: | MODEL: | | YEAR: | |
|-------|--------|---|-------|---|
| DATE: | ODOMETER: START \| END | | TOTAL: | DESTINATION / PURPOSE: |
| | | | | |
| | | | | |
| | | | | |
| | | | | |
| | | | | |
| | | | | |
| | | | | |
| | | | | |
| | | | | |
| | | | | |
| | | | | |
| | | | | |
| | | | | |

# MILEAGE LOG

| MAKE: | | MODEL: | | YEAR: |
|---|---|---|---|---|
| **DATE:** | **ODOMETER:** **START \| END** | | **TOTAL:** | **DESTINATION / PURPOSE:** |
| | | | | |
| | | | | |
| | | | | |
| | | | | |
| | | | | |
| | | | | |
| | | | | |
| | | | | |
| | | | | |
| | | | | |
| | | | | |
| | | | | |
| | | | | |
| | | | | |

# MILEAGE LOG

| MAKE: | | | MODEL: | | YEAR: | |
|---|---|---|---|---|---|---|

| DATE: | ODOMETER:<br>START \| END | | TOTAL: | DESTINATION /<br>PURPOSE: |
|---|---|---|---|---|
| | | | | |
| | | | | |
| | | | | |
| | | | | |
| | | | | |
| | | | | |
| | | | | |
| | | | | |
| | | | | |
| | | | | |
| | | | | |
| | | | | |
| | | | | |
| | | | | |

# MILEAGE LOG

| MAKE: | | MODEL: | | YEAR: |
|---|---|---|---|---|
| **DATE:** | **ODOMETER:** <br> **START \| END** | | **TOTAL:** | **DESTINATION /** <br> **PURPOSE:** |
| | | | | |
| | | | | |
| | | | | |
| | | | | |
| | | | | |
| | | | | |
| | | | | |
| | | | | |
| | | | | |
| | | | | |
| | | | | |
| | | | | |
| | | | | |
| | | | | |

# MILEAGE LOG

| MAKE: | | MODEL: | | YEAR: |
|---|---|---|---|---|
| DATE: | ODOMETER:<br>START \| END | | TOTAL: | DESTINATION /<br>PURPOSE: |
| | | | | |
| | | | | |
| | | | | |
| | | | | |
| | | | | |
| | | | | |
| | | | | |
| | | | | |
| | | | | |
| | | | | |
| | | | | |
| | | | | |
| | | | | |
| | | | | |

# MILEAGE LOG

| MAKE: | | MODEL: | | YEAR: |
|---|---|---|---|---|

| DATE: | ODOMETER: START \| END | | TOTAL: | DESTINATION / PURPOSE: |
|---|---|---|---|---|
| | | | | |
| | | | | |
| | | | | |
| | | | | |
| | | | | |
| | | | | |
| | | | | |
| | | | | |
| | | | | |
| | | | | |
| | | | | |
| | | | | |
| | | | | |
| | | | | |

# MILEAGE LOG

| MAKE: | | MODEL: | | YEAR: |
|-------|---|--------|--------|-------|
| **DATE:** | **ODOMETER:**<br>**START \| END** | | **TOTAL:** | **DESTINATION /**<br>**PURPOSE:** |
| | | | | |
| | | | | |
| | | | | |
| | | | | |
| | | | | |
| | | | | |
| | | | | |
| | | | | |
| | | | | |
| | | | | |
| | | | | |
| | | | | |
| | | | | |
| | | | | |

# MILEAGE LOG

| DATE: | ODOMETER:<br>START \| END | | TOTAL: | DESTINATION /<br>PURPOSE: |
|---|---|---|---|---|
| **MAKE:** | | **MODEL:** | | **YEAR:** |
| | | | | |
| | | | | |
| | | | | |
| | | | | |
| | | | | |
| | | | | |
| | | | | |
| | | | | |
| | | | | |
| | | | | |
| | | | | |
| | | | | |
| | | | | |
| | | | | |

# MILEAGE LOG

| MAKE: | | MODEL: | | YEAR: |
|---|---|---|---|---|
| DATE: | ODOMETER: START \| END | | TOTAL: | DESTINATION / PURPOSE: |
| | | | | |
| | | | | |
| | | | | |
| | | | | |
| | | | | |
| | | | | |
| | | | | |
| | | | | |
| | | | | |
| | | | | |
| | | | | |
| | | | | |
| | | | | |
| | | | | |

# MILEAGE LOG

| DATE: | ODOMETER: START \| END | | TOTAL: | DESTINATION / PURPOSE: |
|-------|------------|------|--------|------------------------|
| **MAKE:** | **MODEL:** | | **YEAR:** | |
| | | | | |
| | | | | |
| | | | | |
| | | | | |
| | | | | |
| | | | | |
| | | | | |
| | | | | |
| | | | | |
| | | | | |
| | | | | |
| | | | | |
| | | | | |
| | | | | |

# MILEAGE LOG

| MAKE: | | MODEL: | | YEAR: |
|---|---|---|---|---|
| DATE: | ODOMETER:<br>START \| END | | TOTAL: | DESTINATION /<br>PURPOSE: |
| | | | | |
| | | | | |
| | | | | |
| | | | | |
| | | | | |
| | | | | |
| | | | | |
| | | | | |
| | | | | |
| | | | | |
| | | | | |
| | | | | |
| | | | | |
| | | | | |

# MILEAGE LOG

| MAKE: | | MODEL: | | YEAR: |
|---|---|---|---|---|
| DATE: | ODOMETER: START \| END | | TOTAL: | DESTINATION / PURPOSE: |
| | | | | |
| | | | | |
| | | | | |
| | | | | |
| | | | | |
| | | | | |
| | | | | |
| | | | | |
| | | | | |
| | | | | |
| | | | | |
| | | | | |
| | | | | |
| | | | | |

# MILEAGE LOG

| MAKE: | | MODEL: | | YEAR: |
|---|---|---|---|---|
| DATE: | ODOMETER:<br>START \| END | | TOTAL: | DESTINATION /<br>PURPOSE: |
| | | | | |
| | | | | |
| | | | | |
| | | | | |
| | | | | |
| | | | | |
| | | | | |
| | | | | |
| | | | | |
| | | | | |
| | | | | |
| | | | | |
| | | | | |

# MILEAGE LOG

| MAKE: | | MODEL: | | YEAR: |
|---|---|---|---|---|
| **DATE:** | **ODOMETER:**<br>**START \| END** | | **TOTAL:** | **DESTINATION /**<br>**PURPOSE:** |
| | | | | |
| | | | | |
| | | | | |
| | | | | |
| | | | | |
| | | | | |
| | | | | |
| | | | | |
| | | | | |
| | | | | |
| | | | | |
| | | | | |
| | | | | |
| | | | | |
| | | | | |
| | | | | |

# MILEAGE LOG

| MAKE: | | MODEL: | | YEAR: |
|-------|--|--------|--|-------|
| DATE: | ODOMETER:<br>START \| END | | TOTAL: | DESTINATION /<br>PURPOSE: |
| | | | | |
| | | | | |
| | | | | |
| | | | | |
| | | | | |
| | | | | |
| | | | | |
| | | | | |
| | | | | |
| | | | | |
| | | | | |
| | | | | |
| | | | | |
| | | | | |

# MILEAGE LOG

| MAKE: | | MODEL: | | YEAR: |
|---|---|---|---|---|
| DATE: | ODOMETER: START \| END | | TOTAL: | DESTINATION / PURPOSE: |
| | | | | |
| | | | | |
| | | | | |
| | | | | |
| | | | | |
| | | | | |
| | | | | |
| | | | | |
| | | | | |
| | | | | |
| | | | | |
| | | | | |
| | | | | |
| | | | | |

# MILEAGE LOG

| MAKE: | | MODEL: | | YEAR: |
|---|---|---|---|---|
| DATE: | ODOMETER:<br>START \| END | | TOTAL: | DESTINATION /<br>PURPOSE: |
| | | | | |
| | | | | |
| | | | | |
| | | | | |
| | | | | |
| | | | | |
| | | | | |
| | | | | |
| | | | | |
| | | | | |
| | | | | |
| | | | | |
| | | | | |

# MILEAGE LOG

| MAKE: | | MODEL: | | YEAR: |
|-------|---|--------|---|-------|
| **DATE:** | **ODOMETER:**<br>**START \| END** | | **TOTAL:** | **DESTINATION /**<br>**PURPOSE:** |
| | | | | |
| | | | | |
| | | | | |
| | | | | |
| | | | | |
| | | | | |
| | | | | |
| | | | | |
| | | | | |
| | | | | |
| | | | | |
| | | | | |
| | | | | |
| | | | | |
| | | | | |

# MILEAGE LOG

| MAKE: | | MODEL: | | YEAR: |
|---|---|---|---|---|
| **DATE:** | **ODOMETER:**<br>**START \| END** | | **TOTAL:** | **DESTINATION /**<br>**PURPOSE:** |
| | | | | |
| | | | | |
| | | | | |
| | | | | |
| | | | | |
| | | | | |
| | | | | |
| | | | | |
| | | | | |
| | | | | |
| | | | | |
| | | | | |
| | | | | |
| | | | | |

# MILEAGE LOG

| MAKE: | | MODEL: | | YEAR: |
|---|---|---|---|---|
| **DATE:** | **ODOMETER: START \| END** | | **TOTAL:** | **DESTINATION / PURPOSE:** |
| | | | | |
| | | | | |
| | | | | |
| | | | | |
| | | | | |
| | | | | |
| | | | | |
| | | | | |
| | | | | |
| | | | | |
| | | | | |
| | | | | |
| | | | | |
| | | | | |
| | | | | |

# MILEAGE LOG

| MAKE: | | MODEL: | | YEAR: |
|---|---|---|---|---|
| DATE: | ODOMETER: START \| END | | TOTAL: | DESTINATION / PURPOSE: |
| | | | | |
| | | | | |
| | | | | |
| | | | | |
| | | | | |
| | | | | |
| | | | | |
| | | | | |
| | | | | |
| | | | | |
| | | | | |
| | | | | |
| | | | | |
| | | | | |

# MILEAGE LOG

| MAKE: | | MODEL: | | YEAR: |
|---|---|---|---|---|
| **DATE:** | **ODOMETER:** **START \| END** | | **TOTAL:** | **DESTINATION /** **PURPOSE:** |
| | | | | |
| | | | | |
| | | | | |
| | | | | |
| | | | | |
| | | | | |
| | | | | |
| | | | | |
| | | | | |
| | | | | |
| | | | | |
| | | | | |
| | | | | |

# MILEAGE LOG

| MAKE: | | MODEL: | | YEAR: |
|-------|---|--------|---|-------|
| DATE: | ODOMETER:<br>START \| END | | TOTAL: | DESTINATION /<br>PURPOSE: |
| | | | | |
| | | | | |
| | | | | |
| | | | | |
| | | | | |
| | | | | |
| | | | | |
| | | | | |
| | | | | |
| | | | | |
| | | | | |
| | | | | |
| | | | | |
| | | | | |

# MILEAGE LOG

| DATE: | ODOMETER: START \| END | | TOTAL: | DESTINATION / PURPOSE: |
|-------|---------|---------|--------|------------------------|
| MAKE: | | MODEL: | | YEAR: |
| | | | | |
| | | | | |
| | | | | |
| | | | | |
| | | | | |
| | | | | |
| | | | | |
| | | | | |
| | | | | |
| | | | | |
| | | | | |
| | | | | |
| | | | | |
| | | | | |

# MILEAGE LOG

| MAKE: | MODEL: | | YEAR: |
|-------|--------|-------|-------|
| DATE: | ODOMETER:<br>START \| END | TOTAL: | DESTINATION /<br>PURPOSE: |
| | | | |
| | | | |
| | | | |
| | | | |
| | | | |
| | | | |
| | | | |
| | | | |
| | | | |
| | | | |
| | | | |
| | | | |
| | | | |
| | | | |

# MILEAGE LOG

| MAKE: | | MODEL: | | YEAR: |
|---|---|---|---|---|
| DATE: | ODOMETER: START \| END | | TOTAL: | DESTINATION / PURPOSE: |
| | | | | |
| | | | | |
| | | | | |
| | | | | |
| | | | | |
| | | | | |
| | | | | |
| | | | | |
| | | | | |
| | | | | |
| | | | | |
| | | | | |
| | | | | |
| | | | | |

# MILEAGE LOG

| MAKE: | | MODEL: | | YEAR: |
|---|---|---|---|---|
| DATE: | ODOMETER:<br>START \| END | | TOTAL: | DESTINATION /<br>PURPOSE: |
| | | | | |
| | | | | |
| | | | | |
| | | | | |
| | | | | |
| | | | | |
| | | | | |
| | | | | |
| | | | | |
| | | | | |
| | | | | |
| | | | | |
| | | | | |
| | | | | |

# MILEAGE LOG

| MAKE: | | MODEL: | | YEAR: |
|---|---|---|---|---|
| **DATE:** | **ODOMETER:** **START \| END** | | **TOTAL:** | **DESTINATION / PURPOSE:** |
| | | | | |
| | | | | |
| | | | | |
| | | | | |
| | | | | |
| | | | | |
| | | | | |
| | | | | |
| | | | | |
| | | | | |
| | | | | |
| | | | | |
| | | | | |
| | | | | |

# MILEAGE LOG

| MAKE: | | MODEL: | | YEAR: |
|---|---|---|---|---|
| **DATE:** | **ODOMETER:**<br>**START \| END** | | **TOTAL:** | **DESTINATION /**<br>**PURPOSE:** |
| | | | | |
| | | | | |
| | | | | |
| | | | | |
| | | | | |
| | | | | |
| | | | | |
| | | | | |
| | | | | |
| | | | | |
| | | | | |
| | | | | |
| | | | | |
| | | | | |

# MILEAGE LOG

| MAKE: | | MODEL: | | YEAR: |
|---|---|---|---|---|
| **DATE:** | **ODOMETER:**<br>**START \| END** | | **TOTAL:** | **DESTINATION /**<br>**PURPOSE:** |
| | | | | |
| | | | | |
| | | | | |
| | | | | |
| | | | | |
| | | | | |
| | | | | |
| | | | | |
| | | | | |
| | | | | |
| | | | | |
| | | | | |
| | | | | |

# MILEAGE LOG

| MAKE: | | MODEL: | | YEAR: |
|---|---|---|---|---|
| **DATE:** | **ODOMETER:** **START \| END** | | **TOTAL:** | **DESTINATION /** **PURPOSE:** |
| | | | | |
| | | | | |
| | | | | |
| | | | | |
| | | | | |
| | | | | |
| | | | | |
| | | | | |
| | | | | |
| | | | | |
| | | | | |
| | | | | |
| | | | | |
| | | | | |

# MILEAGE LOG

| MAKE: | | MODEL: | | YEAR: |
|-------|--|--------|--|-------|
| **DATE:** | **ODOMETER:** **START \| END** | | **TOTAL:** | **DESTINATION /** **PURPOSE:** |
| | | | | |
| | | | | |
| | | | | |
| | | | | |
| | | | | |
| | | | | |
| | | | | |
| | | | | |
| | | | | |
| | | | | |
| | | | | |
| | | | | |
| | | | | |

# MILEAGE LOG

| MAKE: | | MODEL: | | YEAR: |
|---|---|---|---|---|
| DATE: | ODOMETER:<br>START \| END | | TOTAL: | DESTINATION /<br>PURPOSE: |
| | | | | |
| | | | | |
| | | | | |
| | | | | |
| | | | | |
| | | | | |
| | | | | |
| | | | | |
| | | | | |
| | | | | |
| | | | | |
| | | | | |
| | | | | |
| | | | | |
| | | | | |

# MILEAGE LOG

| MAKE: | | MODEL: | | YEAR: |
|---|---|---|---|---|
| **DATE:** | **ODOMETER:** **START \| END** | | **TOTAL:** | **DESTINATION /** **PURPOSE:** |
| | | | | |
| | | | | |
| | | | | |
| | | | | |
| | | | | |
| | | | | |
| | | | | |
| | | | | |
| | | | | |
| | | | | |
| | | | | |
| | | | | |
| | | | | |
| | | | | |
| | | | | |

# MILEAGE LOG

| MAKE: | | MODEL: | | YEAR: |
|---|---|---|---|---|
| DATE: | ODOMETER:<br>START \| END | | TOTAL: | DESTINATION /<br>PURPOSE: |
| | | | | |
| | | | | |
| | | | | |
| | | | | |
| | | | | |
| | | | | |
| | | | | |
| | | | | |
| | | | | |
| | | | | |
| | | | | |
| | | | | |
| | | | | |
| | | | | |

# MILEAGE LOG

| MAKE: | | MODEL: | | YEAR: |
|---|---|---|---|---|
| **DATE:** | **ODOMETER:**<br>**START \| END** | | **TOTAL:** | **DESTINATION /**<br>**PURPOSE:** |
| | | | | |
| | | | | |
| | | | | |
| | | | | |
| | | | | |
| | | | | |
| | | | | |
| | | | | |
| | | | | |
| | | | | |
| | | | | |
| | | | | |
| | | | | |
| | | | | |

# MILEAGE LOG

| MAKE: | | MODEL: | | YEAR: |
|---|---|---|---|---|
| DATE: | ODOMETER:<br>START \| END | | TOTAL: | DESTINATION /<br>PURPOSE: |
| | | | | |
| | | | | |
| | | | | |
| | | | | |
| | | | | |
| | | | | |
| | | | | |
| | | | | |
| | | | | |
| | | | | |
| | | | | |
| | | | | |
| | | | | |
| | | | | |

# MILEAGE LOG

| MAKE: | | MODEL: | | YEAR: |
|---|---|---|---|---|
| **DATE:** | **ODOMETER:** **START \| END** | | **TOTAL:** | **DESTINATION /** **PURPOSE:** |
| | | | | |
| | | | | |
| | | | | |
| | | | | |
| | | | | |
| | | | | |
| | | | | |
| | | | | |
| | | | | |
| | | | | |
| | | | | |
| | | | | |
| | | | | |

# MILEAGE LOG

| MAKE: | | MODEL: | | YEAR: |
|---|---|---|---|---|
| DATE: | ODOMETER:<br>START \| END | | TOTAL: | DESTINATION /<br>PURPOSE: |
| | | | | |
| | | | | |
| | | | | |
| | | | | |
| | | | | |
| | | | | |
| | | | | |
| | | | | |
| | | | | |
| | | | | |
| | | | | |
| | | | | |
| | | | | |
| | | | | |

# MILEAGE LOG

| MAKE: | | MODEL: | | YEAR: |
|---|---|---|---|---|
| **DATE:** | **ODOMETER:**<br>**START \| END** | | **TOTAL:** | **DESTINATION /**<br>**PURPOSE:** |
| | | | | |
| | | | | |
| | | | | |
| | | | | |
| | | | | |
| | | | | |
| | | | | |
| | | | | |
| | | | | |
| | | | | |
| | | | | |
| | | | | |
| | | | | |

# MILEAGE LOG

| MAKE: | | MODEL: | | YEAR: |
|-------|---|--------|---|-------|
| **DATE:** | **ODOMETER:** **START \| END** | | **TOTAL:** | **DESTINATION /** **PURPOSE:** |
| | | | | |
| | | | | |
| | | | | |
| | | | | |
| | | | | |
| | | | | |
| | | | | |
| | | | | |
| | | | | |
| | | | | |
| | | | | |
| | | | | |
| | | | | |
| | | | | |

# MILEAGE LOG

| MAKE: | | MODEL: | | YEAR: |
|-------|--|--------|--|-------|

| DATE: | ODOMETER: START \| END | | TOTAL: | DESTINATION / PURPOSE: |
|-------|---------|--|--------|------------------------|
| | | | | |
| | | | | |
| | | | | |
| | | | | |
| | | | | |
| | | | | |
| | | | | |
| | | | | |
| | | | | |
| | | | | |
| | | | | |
| | | | | |
| | | | | |
| | | | | |
| | | | | |

# MILEAGE LOG

| MAKE: | | MODEL: | | YEAR: |
|---|---|---|---|---|
| DATE: | ODOMETER:<br>START \| END | | TOTAL: | DESTINATION /<br>PURPOSE: |
| | | | | |
| | | | | |
| | | | | |
| | | | | |
| | | | | |
| | | | | |
| | | | | |
| | | | | |
| | | | | |
| | | | | |
| | | | | |
| | | | | |
| | | | | |
| | | | | |

# MILEAGE LOG

| MAKE: | | MODEL: | | YEAR: |
|---|---|---|---|---|
| **DATE:** | **ODOMETER:** **START \| END** | | **TOTAL:** | **DESTINATION /** **PURPOSE:** |
| | | | | |
| | | | | |
| | | | | |
| | | | | |
| | | | | |
| | | | | |
| | | | | |
| | | | | |
| | | | | |
| | | | | |
| | | | | |
| | | | | |
| | | | | |
| | | | | |
| | | | | |
| | | | | |

# MILEAGE LOG

| MAKE: | | MODEL: | | YEAR: |
|---|---|---|---|---|
| **DATE:** | **ODOMETER:**<br>**START \| END** | | **TOTAL:** | **DESTINATION /**<br>**PURPOSE:** |
| | | | | |
| | | | | |
| | | | | |
| | | | | |
| | | | | |
| | | | | |
| | | | | |
| | | | | |
| | | | | |
| | | | | |
| | | | | |
| | | | | |
| | | | | |
| | | | | |

# MILEAGE LOG

| MAKE: | | MODEL: | | YEAR: |
|-------|--|--------|--|-------|
| **DATE:** | **ODOMETER:**<br>**START \| END** | | **TOTAL:** | **DESTINATION /**<br>**PURPOSE:** |
| | | | | |
| | | | | |
| | | | | |
| | | | | |
| | | | | |
| | | | | |
| | | | | |
| | | | | |
| | | | | |
| | | | | |
| | | | | |
| | | | | |
| | | | | |
| | | | | |

# MILEAGE LOG

| MAKE: | | MODEL: | | YEAR: |
|---|---|---|---|---|
| DATE: | ODOMETER:<br>START \| END | | TOTAL: | DESTINATION /<br>PURPOSE: |
| | | | | |
| | | | | |
| | | | | |
| | | | | |
| | | | | |
| | | | | |
| | | | | |
| | | | | |
| | | | | |
| | | | | |
| | | | | |
| | | | | |
| | | | | |
| | | | | |

# MILEAGE LOG

| MAKE: | | MODEL: | | YEAR: |
|---|---|---|---|---|
| **DATE:** | **ODOMETER:**<br>**START \| END** | | **TOTAL:** | **DESTINATION /**<br>**PURPOSE:** |
| | | | | |
| | | | | |
| | | | | |
| | | | | |
| | | | | |
| | | | | |
| | | | | |
| | | | | |
| | | | | |
| | | | | |
| | | | | |
| | | | | |
| | | | | |
| | | | | |
| | | | | |

# MILEAGE LOG

| MAKE: | | MODEL: | | YEAR: |
|---|---|---|---|---|
| **DATE:** | **ODOMETER:** **START \| END** | | **TOTAL:** | **DESTINATION /** **PURPOSE:** |
| | | | | |
| | | | | |
| | | | | |
| | | | | |
| | | | | |
| | | | | |
| | | | | |
| | | | | |
| | | | | |
| | | | | |
| | | | | |
| | | | | |
| | | | | |
| | | | | |

# MILEAGE LOG

| MAKE: | | MODEL: | | YEAR: |
|---|---|---|---|---|
| **DATE:** | **ODOMETER:**<br>**START \| END** | | **TOTAL:** | **DESTINATION /**<br>**PURPOSE:** |
| | | | | |
| | | | | |
| | | | | |
| | | | | |
| | | | | |
| | | | | |
| | | | | |
| | | | | |
| | | | | |
| | | | | |
| | | | | |
| | | | | |
| | | | | |
| | | | | |

# MILEAGE LOG

| MAKE: | | MODEL: | | YEAR: |
|---|---|---|---|---|
| **DATE:** | **ODOMETER:** **START \| END** | | **TOTAL:** | **DESTINATION /** **PURPOSE:** |
| | | | | |
| | | | | |
| | | | | |
| | | | | |
| | | | | |
| | | | | |
| | | | | |
| | | | | |
| | | | | |
| | | | | |
| | | | | |
| | | | | |
| | | | | |
| | | | | |

# MILEAGE LOG

| MAKE: | MODEL: | | YEAR: | |
|---|---|---|---|---|

| DATE: | ODOMETER: START \| END | | TOTAL: | DESTINATION / PURPOSE: |
|---|---|---|---|---|
| | | | | |
| | | | | |
| | | | | |
| | | | | |
| | | | | |
| | | | | |
| | | | | |
| | | | | |
| | | | | |
| | | | | |
| | | | | |
| | | | | |
| | | | | |
| | | | | |

# MILEAGE LOG

| MAKE: | MODEL: | | YEAR: |
|-------|--------|--------|-------|
| DATE: | ODOMETER:<br>START \| END | TOTAL: | DESTINATION /<br>PURPOSE: |
| | | | |
| | | | |
| | | | |
| | | | |
| | | | |
| | | | |
| | | | |
| | | | |
| | | | |
| | | | |
| | | | |
| | | | |
| | | | |
| | | | |

# MILEAGE LOG

| MAKE: | | MODEL: | | YEAR: |
|-------|--|--------|--|-------|
| DATE: | ODOMETER:<br>START \| END | | TOTAL: | DESTINATION /<br>PURPOSE: |
| | | | | |
| | | | | |
| | | | | |
| | | | | |
| | | | | |
| | | | | |
| | | | | |
| | | | | |
| | | | | |
| | | | | |
| | | | | |
| | | | | |
| | | | | |
| | | | | |

# MILEAGE LOG

| MAKE: | | MODEL: | | YEAR: |
|---|---|---|---|---|
| DATE: | ODOMETER:<br>START \| END | | TOTAL: | DESTINATION /<br>PURPOSE: |
| | | | | |
| | | | | |
| | | | | |
| | | | | |
| | | | | |
| | | | | |
| | | | | |
| | | | | |
| | | | | |
| | | | | |
| | | | | |
| | | | | |
| | | | | |
| | | | | |

# MILEAGE LOG

| MAKE: | | MODEL: | | YEAR: |
|---|---|---|---|---|
| **DATE:** | **ODOMETER:**<br>**START \| END** | | **TOTAL:** | **DESTINATION /**<br>**PURPOSE:** |
| | | | | |
| | | | | |
| | | | | |
| | | | | |
| | | | | |
| | | | | |
| | | | | |
| | | | | |
| | | | | |
| | | | | |
| | | | | |
| | | | | |
| | | | | |
| | | | | |

# MILEAGE LOG

| MAKE: | | MODEL: | | YEAR: |
|---|---|---|---|---|
| **DATE:** | **ODOMETER:** **START \| END** | | **TOTAL:** | **DESTINATION /** **PURPOSE:** |
| | | | | |
| | | | | |
| | | | | |
| | | | | |
| | | | | |
| | | | | |
| | | | | |
| | | | | |
| | | | | |
| | | | | |
| | | | | |
| | | | | |
| | | | | |
| | | | | |

# MILEAGE LOG

| MAKE: | | MODEL: | | YEAR: |
|---|---|---|---|---|
| **DATE:** | **ODOMETER:** START \| END | | **TOTAL:** | **DESTINATION / PURPOSE:** |
| | | | | |
| | | | | |
| | | | | |
| | | | | |
| | | | | |
| | | | | |
| | | | | |
| | | | | |
| | | | | |
| | | | | |
| | | | | |
| | | | | |
| | | | | |
| | | | | |

# MILEAGE LOG

| MAKE: | | MODEL: | | YEAR: |
|-------|--|--------|--|-------|
| **DATE:** | **ODOMETER:** <br> **START \| END** | | **TOTAL:** | **DESTINATION /** <br> **PURPOSE:** |
| | | | | |
| | | | | |
| | | | | |
| | | | | |
| | | | | |
| | | | | |
| | | | | |
| | | | | |
| | | | | |
| | | | | |
| | | | | |
| | | | | |
| | | | | |
| | | | | |

# MILEAGE LOG

| MAKE: | | MODEL: | | YEAR: |
|---|---|---|---|---|
| DATE: | ODOMETER:<br>START \| END | | TOTAL: | DESTINATION /<br>PURPOSE: |
| | | | | |
| | | | | |
| | | | | |
| | | | | |
| | | | | |
| | | | | |
| | | | | |
| | | | | |
| | | | | |
| | | | | |
| | | | | |
| | | | | |
| | | | | |
| | | | | |
| | | | | |
| | | | | |

# MILEAGE LOG

| MAKE: | | MODEL: | | YEAR: |
|---|---|---|---|---|
| **DATE:** | **ODOMETER:**<br>**START \| END** | | **TOTAL:** | **DESTINATION /**<br>**PURPOSE:** |
| | | | | |
| | | | | |
| | | | | |
| | | | | |
| | | | | |
| | | | | |
| | | | | |
| | | | | |
| | | | | |
| | | | | |
| | | | | |
| | | | | |
| | | | | |
| | | | | |

# MILEAGE LOG

| MAKE: | | MODEL: | | YEAR: |
|---|---|---|---|---|
| **DATE:** | **ODOMETER:**<br>**START \| END** | | **TOTAL:** | **DESTINATION /**<br>**PURPOSE:** |
| | | | | |
| | | | | |
| | | | | |
| | | | | |
| | | | | |
| | | | | |
| | | | | |
| | | | | |
| | | | | |
| | | | | |
| | | | | |
| | | | | |
| | | | | |
| | | | | |

# MILEAGE LOG

| DATE: | ODOMETER:<br>START \| END | | TOTAL: | DESTINATION /<br>PURPOSE: |
|-------|-------|-------|--------|---------------------------|
| **MAKE:** | | **MODEL:** | | **YEAR:** |
| | | | | |
| | | | | |
| | | | | |
| | | | | |
| | | | | |
| | | | | |
| | | | | |
| | | | | |
| | | | | |
| | | | | |
| | | | | |
| | | | | |
| | | | | |
| | | | | |

# MILEAGE LOG

| MAKE: | | MODEL: | | YEAR: |
|---|---|---|---|---|
| **DATE:** | **ODOMETER:**<br>**START \| END** | | **TOTAL:** | **DESTINATION /**<br>**PURPOSE:** |
| | | | | |
| | | | | |
| | | | | |
| | | | | |
| | | | | |
| | | | | |
| | | | | |
| | | | | |
| | | | | |
| | | | | |
| | | | | |
| | | | | |
| | | | | |

# MILEAGE LOG

| MAKE: | | MODEL: | | YEAR: |
|---|---|---|---|---|
| **DATE:** | **ODOMETER:** **START \| END** | | **TOTAL:** | **DESTINATION / PURPOSE:** |
| | | | | |
| | | | | |
| | | | | |
| | | | | |
| | | | | |
| | | | | |
| | | | | |
| | | | | |
| | | | | |
| | | | | |
| | | | | |
| | | | | |
| | | | | |
| | | | | |

# MILEAGE LOG

| MAKE: | | MODEL: | | YEAR: |
|-------|--|--------|--|-------|
| **DATE:** | **ODOMETER:**<br>**START \| END** | | **TOTAL:** | **DESTINATION /**<br>**PURPOSE:** |
| | | | | |
| | | | | |
| | | | | |
| | | | | |
| | | | | |
| | | | | |
| | | | | |
| | | | | |
| | | | | |
| | | | | |
| | | | | |
| | | | | |
| | | | | |

# MILEAGE LOG

| MAKE: | MODEL: | | YEAR: |
|---|---|---|---|
| DATE: | ODOMETER:<br>START \| END | TOTAL: | DESTINATION /<br>PURPOSE: |
| | | | |
| | | | |
| | | | |
| | | | |
| | | | |
| | | | |
| | | | |
| | | | |
| | | | |
| | | | |
| | | | |
| | | | |
| | | | |
| | | | |

# MILEAGE LOG

| MAKE: | | MODEL: | | YEAR: |
|---|---|---|---|---|
| **DATE:** | **ODOMETER:** **START \| END** | | **TOTAL:** | **DESTINATION /** **PURPOSE:** |
| | | | | |
| | | | | |
| | | | | |
| | | | | |
| | | | | |
| | | | | |
| | | | | |
| | | | | |
| | | | | |
| | | | | |
| | | | | |
| | | | | |
| | | | | |
| | | | | |

# MILEAGE LOG

| MAKE: | MODEL: | | YEAR: |
|---|---|---|---|
| **DATE:** | **ODOMETER:** **START \| END** | **TOTAL:** | **DESTINATION / PURPOSE:** |
| | | | |
| | | | |
| | | | |
| | | | |
| | | | |
| | | | |
| | | | |
| | | | |
| | | | |
| | | | |
| | | | |
| | | | |
| | | | |
| | | | |

# MILEAGE LOG

| MAKE: | | MODEL: | | YEAR: |
|---|---|---|---|---|
| **DATE:** | **ODOMETER:**<br>**START \| END** | | **TOTAL:** | **DESTINATION /**<br>**PURPOSE:** |
| | | | | |
| | | | | |
| | | | | |
| | | | | |
| | | | | |
| | | | | |
| | | | | |
| | | | | |
| | | | | |
| | | | | |
| | | | | |
| | | | | |
| | | | | |
| | | | | |

# MILEAGE LOG

| MAKE: | | MODEL: | | YEAR: |
|---|---|---|---|---|
| DATE: | ODOMETER: START \| END | | TOTAL: | DESTINATION / PURPOSE: |
| | | | | |
| | | | | |
| | | | | |
| | | | | |
| | | | | |
| | | | | |
| | | | | |
| | | | | |
| | | | | |
| | | | | |
| | | | | |
| | | | | |
| | | | | |
| | | | | |

# MILEAGE LOG

| MAKE: | | MODEL: | | YEAR: |
|-------|---|--------|---|-------|
| **DATE:** | **ODOMETER:** <br> **START \| END** | | **TOTAL:** | **DESTINATION /** <br> **PURPOSE:** |
| | | | | |
| | | | | |
| | | | | |
| | | | | |
| | | | | |
| | | | | |
| | | | | |
| | | | | |
| | | | | |
| | | | | |
| | | | | |
| | | | | |
| | | | | |

# MILEAGE LOG

| DATE: | ODOMETER: START \| END | | TOTAL: | DESTINATION / PURPOSE: |
|---|---|---|---|---|
| **MAKE:** | **MODEL:** | | **YEAR:** | |
| | | | | |
| | | | | |
| | | | | |
| | | | | |
| | | | | |
| | | | | |
| | | | | |
| | | | | |
| | | | | |
| | | | | |
| | | | | |
| | | | | |
| | | | | |
| | | | | |

# MILEAGE LOG

| MAKE: | | MODEL: | | YEAR: |
|---|---|---|---|---|
| **DATE:** | **ODOMETER:**<br>**START \| END** | | **TOTAL:** | **DESTINATION /**<br>**PURPOSE:** |
| | | | | |
| | | | | |
| | | | | |
| | | | | |
| | | | | |
| | | | | |
| | | | | |
| | | | | |
| | | | | |
| | | | | |
| | | | | |
| | | | | |
| | | | | |
| | | | | |

# MILEAGE LOG

| MAKE: | | MODEL: | | YEAR: |
|---|---|---|---|---|
| **DATE:** | **ODOMETER:**<br>**START \| END** | | **TOTAL:** | **DESTINATION /**<br>**PURPOSE:** |
| | | | | |
| | | | | |
| | | | | |
| | | | | |
| | | | | |
| | | | | |
| | | | | |
| | | | | |
| | | | | |
| | | | | |
| | | | | |
| | | | | |
| | | | | |
| | | | | |

# MILEAGE LOG

| MAKE: | | | MODEL: | | YEAR: | |
|---|---|---|---|---|---|---|
| **DATE:** | | **ODOMETER:** <br> **START \| END** | **TOTAL:** | | **DESTINATION /** <br> **PURPOSE:** | |
| | | | | | | |
| | | | | | | |
| | | | | | | |
| | | | | | | |
| | | | | | | |
| | | | | | | |
| | | | | | | |
| | | | | | | |
| | | | | | | |
| | | | | | | |
| | | | | | | |
| | | | | | | |
| | | | | | | |
| | | | | | | |

# MILEAGE LOG

| MAKE: | | MODEL: | | YEAR: |
|---|---|---|---|---|
| **DATE:** | **ODOMETER:** **START \| END** | | **TOTAL:** | **DESTINATION / PURPOSE:** |
| | | | | |
| | | | | |
| | | | | |
| | | | | |
| | | | | |
| | | | | |
| | | | | |
| | | | | |
| | | | | |
| | | | | |
| | | | | |
| | | | | |
| | | | | |
| | | | | |

# MILEAGE LOG

| MAKE: | | MODEL: | | YEAR: |
|---|---|---|---|---|

| DATE: | ODOMETER: START \| END | | TOTAL: | DESTINATION / PURPOSE: |
|---|---|---|---|---|
| | | | | |
| | | | | |
| | | | | |
| | | | | |
| | | | | |
| | | | | |
| | | | | |
| | | | | |
| | | | | |
| | | | | |
| | | | | |
| | | | | |
| | | | | |

# MILEAGE LOG

| MAKE: | MODEL: | | YEAR: | |
|-------|--------|---|-------|---|
| DATE: | ODOMETER: START \| END | | TOTAL: | DESTINATION / PURPOSE: |
| | | | | |
| | | | | |
| | | | | |
| | | | | |
| | | | | |
| | | | | |
| | | | | |
| | | | | |
| | | | | |
| | | | | |
| | | | | |
| | | | | |
| | | | | |
| | | | | |
| | | | | |

# MILEAGE LOG

| MAKE: | | MODEL: | | YEAR: |
|---|---|---|---|---|
| DATE: | ODOMETER:<br>START \| END | | TOTAL: | DESTINATION /<br>PURPOSE: |
| | | | | |
| | | | | |
| | | | | |
| | | | | |
| | | | | |
| | | | | |
| | | | | |
| | | | | |
| | | | | |
| | | | | |
| | | | | |
| | | | | |
| | | | | |
| | | | | |

# MILEAGE LOG

| MAKE: | | MODEL: | | YEAR: |
|---|---|---|---|---|
| **DATE:** | **ODOMETER:**<br>**START \| END** | | **TOTAL:** | **DESTINATION /**<br>**PURPOSE:** |
| | | | | |
| | | | | |
| | | | | |
| | | | | |
| | | | | |
| | | | | |
| | | | | |
| | | | | |
| | | | | |
| | | | | |
| | | | | |
| | | | | |
| | | | | |
| | | | | |

# MILEAGE LOG

| MAKE: | | MODEL: | | YEAR: |
|---|---|---|---|---|
| **DATE:** | **ODOMETER:** START \| END | | **TOTAL:** | **DESTINATION / PURPOSE:** |
| | | | | |
| | | | | |
| | | | | |
| | | | | |
| | | | | |
| | | | | |
| | | | | |
| | | | | |
| | | | | |
| | | | | |
| | | | | |
| | | | | |
| | | | | |
| | | | | |

# MILEAGE LOG

| MAKE: | | MODEL: | | YEAR: |
|---|---|---|---|---|
| DATE: | ODOMETER:<br>START \| END | | TOTAL: | DESTINATION /<br>PURPOSE: |
| | | | | |
| | | | | |
| | | | | |
| | | | | |
| | | | | |
| | | | | |
| | | | | |
| | | | | |
| | | | | |
| | | | | |
| | | | | |
| | | | | |
| | | | | |
| | | | | |

# MILEAGE LOG

| MAKE: | | MODEL: | | YEAR: | |
|-------|---|--------|---|-------|---|
| DATE: | ODOMETER:<br>START \| END | | TOTAL: | DESTINATION /<br>PURPOSE: | |
| | | | | | |
| | | | | | |
| | | | | | |
| | | | | | |
| | | | | | |
| | | | | | |
| | | | | | |
| | | | | | |
| | | | | | |
| | | | | | |
| | | | | | |
| | | | | | |
| | | | | | |
| | | | | | |

# MILEAGE LOG

| MAKE: | | MODEL: | | YEAR: |
|---|---|---|---|---|
| DATE: | ODOMETER: START \| END | | TOTAL: | DESTINATION / PURPOSE: |
| | | | | |
| | | | | |
| | | | | |
| | | | | |
| | | | | |
| | | | | |
| | | | | |
| | | | | |
| | | | | |
| | | | | |
| | | | | |
| | | | | |
| | | | | |
| | | | | |

Made in the USA
Las Vegas, NV
01 July 2021